Topical Conference: Applications of Data Science to Molecules and Materials

AA002623

Held at the 2024 AIChE Annual Meeting

San Diego, California, USA
27-31 October 2024

ISBN: 979-8-3313-1683-9

Printed from e-media with permission by:

Curran Associates, Inc.
57 Morehouse Lane
Red Hook, NY 12571

AIChE
120 Wall Street, FL 23
New York, NY 10005-4020

Phone: (800) 242-4363
Fax: (203) 775-5177

www.aiche.org

Additional copies of this publication are available from:

Curran Associates, Inc.
57 Morehouse Lane
Red Hook, NY 12571 USA
Phone: 845-758-0400
Fax: 845-758-2633
Email: curran@proceedings.com
Web: www.proceedings.com

TABLE OF CONTENTS

APPLICATIONS OF DATA SCIENCE TO MOLECULES AND MATERIALS

435a Successful 'Closed Loop' Materials Discovery: the Power and Challenges of Machine Learning 1
Paulette Clancy

435b Using Machine Learning to Overcome Limitations in Electronic Structure Methodology for Chemical Discovery 2
Heather Kulik

435c Accelerating Nanoporous Materials Discovery for Gas Storage and Separations via Advanced Simulation and Materials Characterization 3
Kaihang Shi

435d Camd.Jl: A Generalized Computer-Aided Molecular Design Framework 4
Pierre Walker, Matthew Cox, Sriteja Mantha, Alejandro Gallegos, Yasemin Basdogan, Zhen-Gang Wang

435e Data-Driven Electrolyte Design for Lithium Metal Batteries 5
Solomon Oyakhire, Yi Cui, Stacey Bent

435f Residual Learning-Based Model for Lithium-Ion Battery End-of-Life Prediction 6
Jaewook Lee, Seyeong Park, Seongmin Heo

435g Machine-Learning Guided Approach for Designing Adeno-Associated Viral Proteins in Therapeutic Delivery 8
Danqing Zhu

Author Index

435a Successful 'Closed Loop' Materials Discovery: the Power and Challenges of Machine Learning

Paulette Clancy

The Materials Genome Initiative was started in 2011 with the prospect of using machine learning to make materials "faster, cheaper, better." A decade or so later, we are finally starting to have the tools to exploit that prospect more effectively. A major factor contributing to the inefficiency of material discovery is the large combinatorial space of materials candidates as well as processing conditions, which are often sparsely observed for a given application. Searches of this space are often formed by expert knowledge and clustered
close to known materials. Exhaustive experimental characterization or first principles calculations are also expensive, invariably leading to small available data sets. As a result, there is a need to develop algorithms that can efficiently search this large parameter space capable of dealing with "tiny" data sets. In this talk, we will introduce our approach to mitigate some of these issues using our in-house PAL 2.0, Bayesian optimization code base, that features a chemistry-informed belief model. A key characteristic of PAL 2.0 is the creation of a physics-based hypothesis using XGBoost and Neural Networks for "feature engineering," which provides a physics-based prior to the Gaussian process model used in Bayesian optimization. Our method picks out the physical descriptors that are most representative of the material domain, making the search independent from expert knowledge. Here, we demonstrate PAL 2.0's use in a "closed-loop" set up with our collaborators at Hopkins Advanced Physics Laboratory to discover high temperature shape memory alloys for space actuation applications. We discuss the use of constraints to help "zero in" to manufacturable solutions in the fewest number of optimization iterations. In the process, we discover new multi-principal element alloys (MPEAs) with dramatically improved targeted properties, in this case alloy toughness.

435b Using Machine Learning to Overcome Limitations in Electronic Structure Methodology for Chemical Discovery

Heather Kulik

Machine learning (ML)-accelerated discovery of materials, i.e., via surrogate models paired with efficient optimization algorithms, holds immense promise to overcome the conventional limitations of computational cost of first-principles electronic structure calculations. Nevertheless, surrogate models inherit the bias of the underlying electronic structure method. In most cases, the electronic structure method of choice is Kohn-Sham density functional theory (DFT), which suffers simultaneously from both self-interaction error or density delocalization error and static correlation error, to varying degrees depending on the density functional approximation. When novel and challenging materials, such as open shell transition metal complexes, is the target of a discovery campaign, few benchmarks are liable to be available for selecting the optimal electronic structure method or DFT functional. Furthermore, investigation of large regions of chemical space (e.g., by varying metal, coordination environment, or oxidation state in a transition metal complex) will likely lead to the conclusion that different electronic structure methods are more suitable for some compounds than others. I will first describe our analysis of the sources of error from the underlying DFT functional. I will describe how we have used consensus from functionals to make more robust predictions of spin crossover behavior, as validated by experimental datasets, and I will describe how we have incorporated both prediction of multi- reference character and DFT consensus to find method-insensitive light harvesting chromophores. Next, I will describe our development of ML-informed density functional models, including a recommender that can identify which DFT functional or parameterization is most predictive to obtain accurate properties of transition metal complexes. I will describe our recent efforts to extend this regress-then-classify strategy to a purely regression-based model that acts in its own right as a density functional. Finally, time permitting, I will describe our efforts in predicting, detecting, and correcting for high multireference character in transition metal complexes as well as some of the datasets we have curated for testing electronic structure methods.

435c Accelerating Nanoporous Materials Discovery for Gas Storage and Separations via Advanced Simulation and Materials Characterization

Kaihang Shi

Porosity is ubiquitous, from naturally formed systems (e.g., shale rocks, wood, and human bones) to synthetic materials (e.g., metal-organic frameworks and molecular cages). The ability of nanoporous materials to adsorb various molecules of interest makes them strong candidates for applications in energy storage, chemical separation, sensing, catalysis, and nano-manufacturing. Finding the top-performing candidates, however, is a daunting task due to the exponentially increasing chemical and structural variables in the materials design space. Physics-based simulations and machine learning (ML) have been proven highly efficient in navigating this virtually infinite space.

In this talk, I will first show how the ML force field significantly improves the accuracy of the grand canonical Monte Carlo (GCMC) simulations for modeling challenging gas adsorption in metal-organic frameworks (MOFs), compared to the classical force field. Then I will present a novel concept of "pore graphs", a compact yet comprehensive way to characterize and represent nanoporous materials. The pore graph representation is flexible and informative to retain any cavity information in a porous structure. We envision this new computational characterization method, coupled with ML, will facilitate the discovery and design of advanced nanoporous materials to address global challenges in energy and sustainability.

435d Camd.Jl: A Generalized Computer-Aided Molecular Design Framework

Pierre Walker, Matthew Cox, Sriteja Mantha, Alejandro Gallegos, Yasemin Basdogan, Zhen-Gang Wang

The increasing need for application-specific molecules has underscored the indispensable role of computer-aided molecular design (CAMD) methodologies. CAMD typically necessitates two integral components: a surrogate model for the prediction of pertinent thermodynamic properties and a molecular design toolkit. Although Machine Learning (ML) approaches have garnered considerable popularity in furnishing such surrogates, they are often constrained by their capacity to fulfill singular design objectives. Conversely, Clapeyron.jl, an advanced open-source thermophysical property estimation library, offers a notably generalized and predictive framework, capable of encompassing a wide array of properties relevant to molecular design endeavors. Capitalizing on the capabilities of this robust library, we introduce CAMD.jl, a CAMD framework adept at tackling diverse molecular design challenges of significance. In this talk, employing a graph-based genetic algorithm methodology, we showcase the versatility of CAMD.jl by its application in various contexts, including:

- Design of non-ozone-depleting refrigerants to satisfy particular operating conditions.
- The development of innovative surfactants aimed at minimizing interfacial tension between aqueous and oil phases.
- The formulation of battery electrolytes employing ionic liquids to optimize conductivity.

This framework holds a great deal of promise to be applicable to other important problems in the community such as carbon capture and pharmaceutical solvent-selection.

435e Data-Driven Electrolyte Design for Lithium Metal Batteries

Solomon Oyakhire, Yi Cui, Stacey Bent

Enhancing Coulombic efficiency (CE) plays a pivotal role in facilitating the adoption of high-energy-density lithium metal batteries. While liquid electrolyte engineering has emerged as a promising strategy for improving CE, its inherent complexity makes performance prediction and electrolyte design challenging. In this two-part presentation, we introduce machine learning methods and workflows that enable the prediction of electrolyte performance, the design of new high-performing electrolytes, and the extraction of important scientific insights.

In the first part, we introduce a novel workflow that combines principles of feature engineering, feature selection, and machine learning model assessment. This workflow allows us to extract insights that guide the design of five new, high-performing electrolytes. Leveraging simple features, such as elemental composition that encodes pertinent physics within the electrolytes, we constructed interpretable models using linear regression, random forest, and bagging techniques. Through the results derived from these interpretable models, we identified crucial electrolyte features that are instrumental in achieving high battery efficiency. One such feature is the atomic fraction of oxygen in the solvent, highlighting the significance of reducing solvent oxygen for achieving high Coulombic efficiency (CE). Equipped with this insight and a few others, we formulated five new electrolyte compositions with fluorine-free solvents, one of which attains a high CE of 99.70%.

In the second part, we employ data segmentation in conjunction with machine learning methods to discern crucial performance descriptors within distinct electrolyte efficiency classes. Through this approach, we made a surprising discovery. Common electrolyte performance descriptors like lithium morphology, ionic conductivity, solid electrolyte interphase chemistry, and lithium-electrolyte reactivity, **do not** explain performance variations in electrolytes beyond a Coulombic Efficiency (CE) of 98%. By utilizing new machine learning model assessment techniques, interpretable machine learning models, correlation analysis, and rigorous spectroscopy and electrochemistry characterizations, we unveil the pivotal role of **galvanic corrosion** in accounting for performance disparities within high CE (>98%) electrolytes.

This work underscores the potential of data-driven approaches in expediting the discovery of high-performance electrolytes for lithium metal batteries.

435f Residual Learning-Based Model for Lithium-Ion Battery End-of-Life Prediction

Jaewook Lee, Seyeong Park, Seongmin Heo

Predicting the lifetime of lithium-ion batteries (LIBs) is critical for ensuring safety and optimizing operational strategies. The end-of-life (EOL) of LIB is commonly defined as the point where the state-of-health (SOH), the current cycle's capacity over nominal capacity, is 80%. C-rate, depth-of-discharge (DOD), and ambient temperature (T_{amb}) are the major factors that determine the EOL of LIBs [1]. Predicting EOL becomes particularly challenging when LIBs are subjected to harsh conditions, e.g., rapid charge and discharge, too high or low , as Li plating can cause 'knee', where capacity starts to decrease rapidly at a certain point under these conditions.

While various public datasets describe LIB's capacity fade [2-3], to the best of our knowledge, no dataset exists that would reflect the variation of all three factors and cycled for EOL due to high experimental costs. Furthermore, due to cell-to-cell variability [4] (i.e. the deviation between cells due to uncontrollable changes during manufacturing stages), cells' EOL would differ in some regions even if they are cycled under the same operating conditions.

In this study, we propose a systematic approach for lifetime prediction of LIBs, where two models are sequentially applied to effectively capture the effects of cell-to-cell variability. To do this, we first generate datasets by simulating the operation of LIBs up to EOL with the cell-to-cell variability considerations. For this purpose, we modify an electrochemical model (EM) simulator called LIONSIMBA [6] by adding Li plating effect [7] (overpotential η) to consider some cells where knees exist. Then, we conduct a large number of cycling tests under various major factors and internal cell parameters. With these datasets, the gaussian process regression (GPR) model is first used to predict the nominal EOL (i.e. EOL without considering cell-to-cell variability). Then, the concept of residual learning [5] is adopted to add a bias to our prediction so that the cell-to-cell variability or minor factors' changes can be captured. We calibrate the nominal EOL by calculating the change in EOL when the cell-to-cell variability exists or minor factors change. Specifically, after calculating EOL bias with refined health indexes (RHIs), e.g., -related factors directly, we present additional neural networks that would calculate RHIs from direct health indexes (DHIs) from early cycles, e.g., voltage (V), current (I), and temperature (T) directly and predict final EOL with only operating conditions and DHIs.

[1] Panchal S, Pierre V, Cancian M, Gross O, Estefanous F and Badawy T. (2023). Development and validation of cycle and calendar aging model for 144Ah NMC/graphite battery at multi temperatures, DODs, and C-rates, SAE Technical Paper.

[2] Severson KA, Attia PM, Jin N, Perkins N, Jiang B, Yang Z, et al. Data-driven prediction of battery cycle life before capacity degradation. Nat. Energy 2019;4:383-91.

[3] Attia PM, Grover A, Jin N, Severson KA, Markov TM, Liao Y-H, et al. Closed-loop optimization of fast-charging protocols for batteries with machine learning. Nature 2020;578:397-402.

[4] Beck D, Dechent P, Junker M, Sauer DU and Dubarry M. Inhomogeneities and cell-to-cell variations in lithium-ion batteries, a review. Energies 2021;14:3276.

[5] He K, Zhang X, Ren S and Sun J. Deep residual learning for image recognition. CVPR;2016.

[6] Torchio M, Magni L, Gopaluni RB, Braatz RD and Raimondo DM. Lionsimba: a matlab framework based on a finite volume model suitable for li-ion battery design, simulation, and control. J. Electrochem. Soc. 2016;163:A1192.

[7] Yang X-G, Leng Y, Zhang G, Ge S and Wang C-Y. Modeling of lithium plating induced aging of lithium-ion batteries: Transition from linear to nonlinear aging. J. Power Sources 2017;360:28-40.

435g Machine-Learning Guided Approach for Designing Adeno-Associated Viral Proteins in Therapeutic Delivery

Danqing Zhu

Gene therapy, the delivery of genetic material to the cells of a patient for therapeutic benefit, has been increasingly successful over the past decade. The most successful gene delivery vectors are based on adeno-associated viruses (AAV). Although these natural viruses are safe and non-pathogenic, they present several barriers that limit their efficacy in delivery as they were not evolved by nature for human therapeutic applications. Directed evolution, a strategy involves the iterative genetic diversification of a molecule to create a gene pool and functional selection to isolate variants with optimal properties, has thereby emerged as a powerful approach for re-evolving AAVs of novel and improved functions. In parallel, recent advances in deep sequencing technologies allow millions of sequences to be assayed and used for training machine learning (ML) models for prediction of protein properties. Using the combination of directed evolution and ML-guided design, we have engineered 'designer' AAV variants with greatly improved packaging, diversity, infection capabilities, and immune-resistant properties. The extension of such integration of new technologies will have broad utility in the development of next-generation novel delivery vectors for therapeutic applications of many diseases in the near future.

To optimize our directed evolution library design, we first developed a ML-based framework to design an AAV5-based capsid library, focusing on a 7-amino acid insertion region at the 3-fold symmetry axis that was identified as a retargeting site for cell-specific entry. Using the trained predictive model, we designed individual libraries that lie on an optimal curve that trades off predicted enrichment with maximal sequence diversity. To judge the utility of our *in-silico* design approach, we synthesized two of the libraries and showed that our ML-library achieves 5-fold higher packaging fitness than the state-of-the-art library, with approximately 10-fold more successful variants after primary brain infection. From this, we demonstrated that 1) individual ML-predicted AAV sequences exhibited experimental packaging capability positively correlated to their predications; 2) synthetic ML-libraries present higher packaging capability while retaining their high diversity of variants. We then transfected our synthetic ML-designed library into packaging cell lines and purified capsid libraries were administered to a) primary brain slices and b) population-wide serum pool. Functional selective pressure was imposed by harvesting immune-resistant variants and cell-specific variants (glial- and neuronal-only cells) after 72 hours. Using PCR-based recovery of variants with deep sequencing, we identified converged variants that successfully transduce human glial or neuronal cells as well as escaped neutralizing antibodies.

In summary, our approach offers a roadmap for designing an AAV library that broadly improves packaging functionality while retaining high diversity. Our computational method can generate as many libraries as desired by leveraging one predictive model. This approach can be harnessed to optimize libraries for downstream selection for properties of therapeutic relevance, including production, infectivity, specificity, and others. In particular, using our ML-guided library design, we generated novel variants with specific targeting and immune-resistant behaviors, which will provide potential therapeutics for disease treatment in the future.

NOTES

NOTES

AUTHOR INDEX

Basdogan, Yasemin 4
Bent, Stacey 5
Clancy, Paulette 1
Cox, Matthew 4
Cui, Yi 5
Gallegos, Alejandro 4
Heo, Seongmin 6
Kulik, Heather 2
Lee, Jaewook 6
Mantha, Sriteja 4
Oyakhire, Solomon 5
Park, Seyeong 6
Shi, Kaihang 3
Walker, Pierre 4
Wang, Zhen-Gang 4
Zhu, Danqing 8